DANGEROUS JOURNEYS

EXPLORING CAVES

BY BETSY RATHBURN

BELLWETHER MEDIA · MINNEAPOLIS, MN

Torque brims with excitement perfect for thrill-seekers of all kinds. Discover daring survival skills, explore uncharted worlds, and marvel at mighty engines and extreme sports. In *Torque* books, anything can happen. Are you ready?

This edition first published in 2023 by Bellwether Media, Inc.

Library of Congress Cataloging-in-Publication Data

Names: Rathburn, Betsy, author.
Title: Exploring caves / by Betsy Rathburn.
Description: Minneapolis, MN : Bellwether Media, Inc., 2023. | Series: Torque. Dangerous journeys | Includes bibliographical references and index. | Audience: Ages 7-12 | Audience: Grades 4-6 | Summary: "Amazing photography accompanies engaging information about cave exploration. The combination of high-interest subject matter and light text is intended for students in grades 3 through 7"–Provided by publisher.
Identifiers: LCCN 2022012965 (print) | LCCN 2022012966 (ebook) | ISBN 9781644877647 (library binding) | ISBN 9781648348808 (paperback) | ISBN 9781648348105 (ebook) Subjects: LCSH: Caves–Juvenile literature.
Classification: LCC GB601.2 .R38 2023 (print) | LCC GB601.2 (ebook) | DDC 551.44/7-dc23
LC record available at https://lccn.loc.gov/2022012965
LC ebook record available at https://lccn.loc.gov/2022012966

Editor: Kieran Downs Designer: Josh Brink

Printed in the United States of America, North Mankato, MN.

TABLE OF CONTENTS

UNDERGROUND RIVER

An explorer walks through a long, dark cave. Suddenly, the path splits into two tunnels. She must choose which way to go.

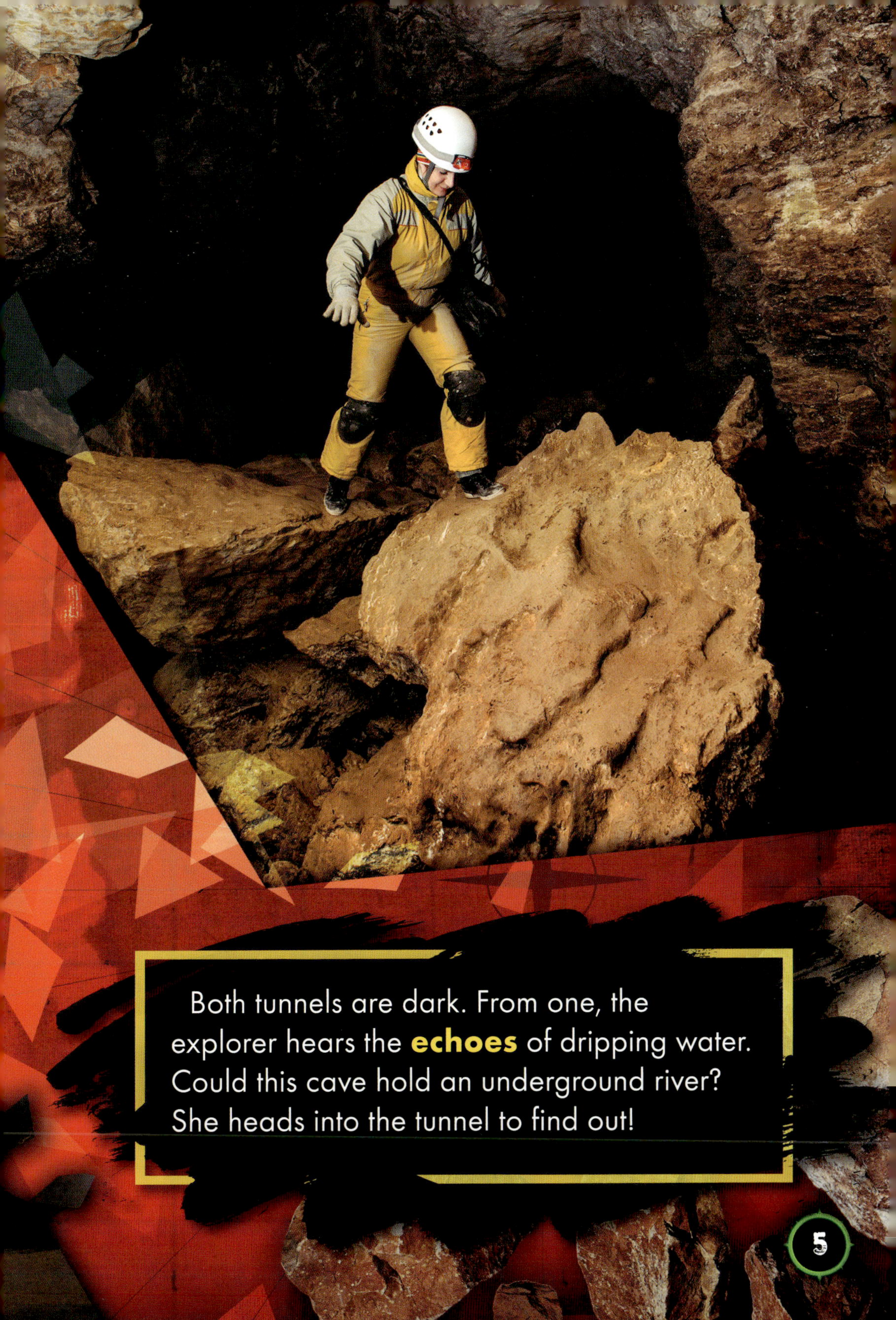

Both tunnels are dark. From one, the explorer hears the **echoes** of dripping water. Could this cave hold an underground river? She heads into the tunnel to find out!

TUNNELS AND CAVERNS

Caves are openings under Earth's surface. They are found on every **continent**! Caves are often found in hillsides or cliffs. Some are found underwater. Others form in ice or in **lava**!

MAMMOTH CAVE

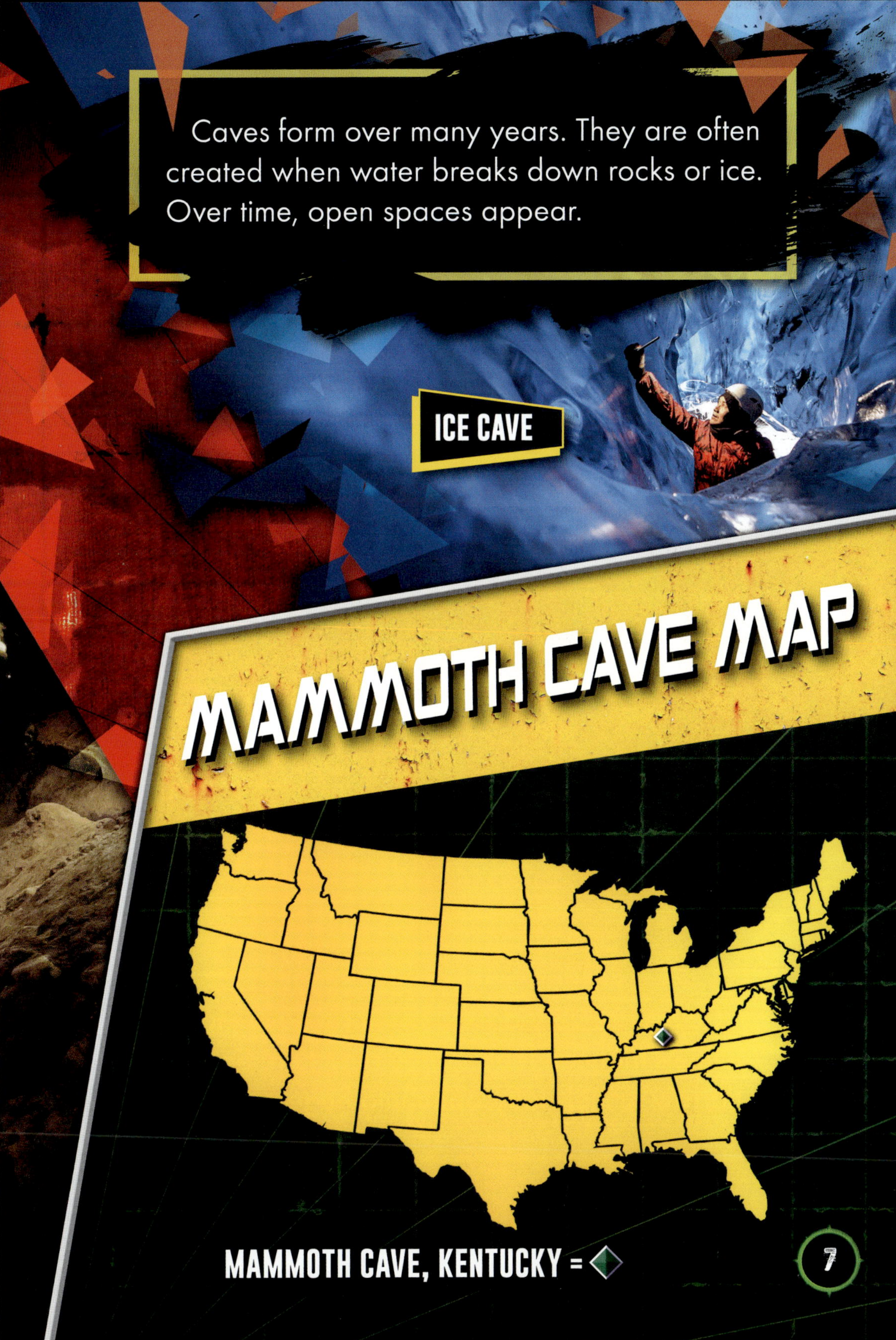

Caves form over many years. They are often created when water breaks down rocks or ice. Over time, open spaces appear.

MAMMOTH CAVE MAP

Caves are very dark. They may have many tunnels that can lead in any direction. Some narrow tunnels may open into huge **caverns**.

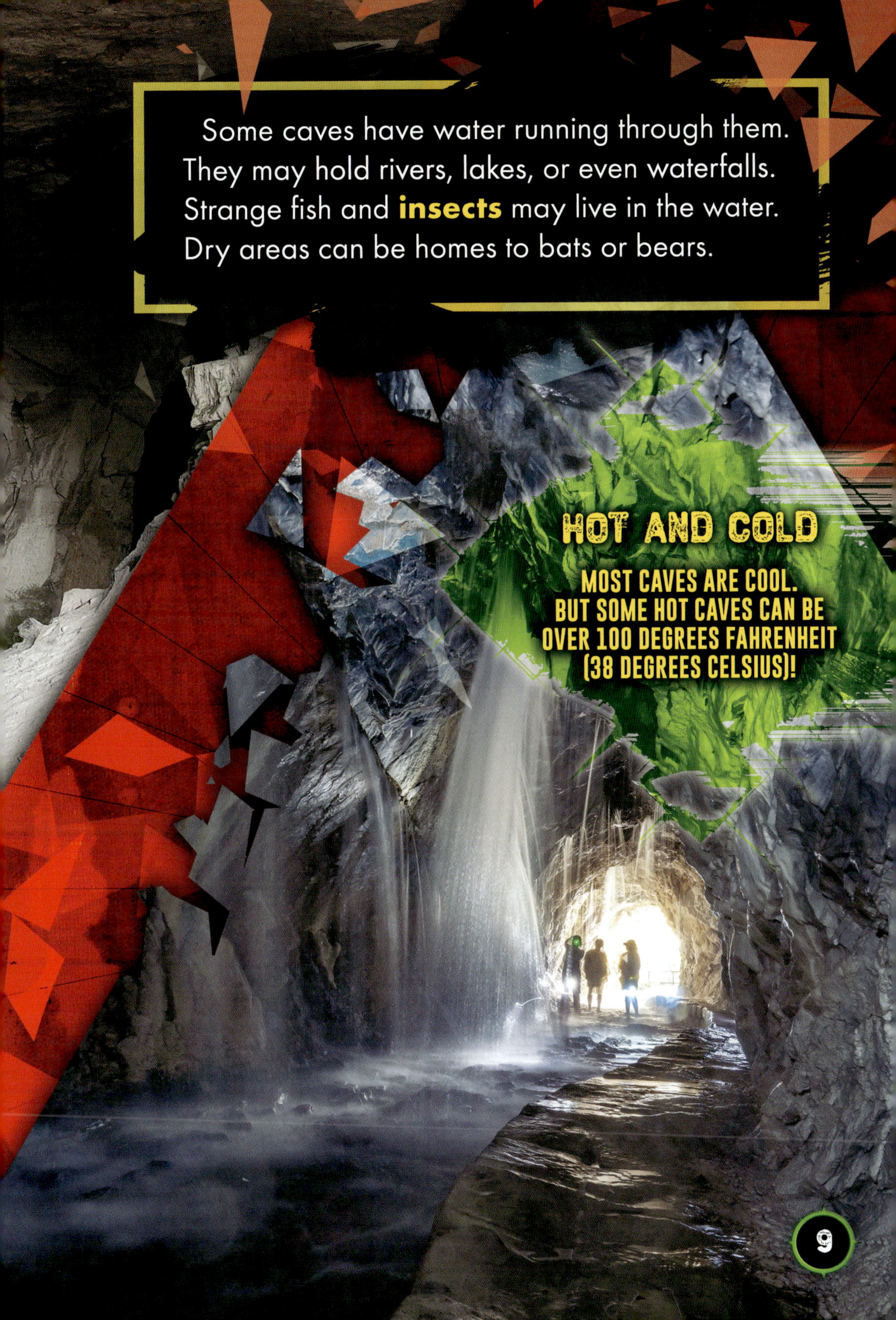

Some caves have water running through them. They may hold rivers, lakes, or even waterfalls. Strange fish and **insects** may live in the water. Dry areas can be homes to bats or bears.

HOT AND COLD

MOST CAVES ARE COOL. BUT SOME HOT CAVES CAN BE OVER 100 DEGREES FAHRENHEIT (38 DEGREES CELSIUS)!

CAVE ART MADE BY EARLY HUMANS

People have used caves for thousands of years. Some of the earliest humans lived in caves. Today, people explore caves to learn about the lives of these early humans.

People also study plants and animals only found in caves. Some people map caves and study how they form. Other people explore caves for fun!

OLM
NOTABLE EXPLORER
NAME: JILL HEINERTH
BORN: JANUARY 16, 1965
JOURNEY: DOVE TO EXPLORE UNDERWATER CAVES IN FLORIDA IN 1998
RESULTS: HELPED CREATE THE FIRST 3D MAP OF AN UNDERWATER CAVE

PLANNING AND PREPARATION

Explorers must prepare before they enter caves. They train for many years. They learn about cave safety. They know the warning signs of dangers such as flooding.

Explorers must also be healthy enough to move through difficult **passages**. They may have to fit into tight spaces. Sometimes they crawl through muddy tunnels.

Underwater cave explorers need to be **certified** in **scuba diving**. They train in smaller caves. Then, they are ready to explore larger caves.

Some caves are very dangerous or **fragile**. Sometimes they can only be explored during certain hours of the day. Cave explorers need **permits** to enter them.

PLANNING YOUR JOURNEY

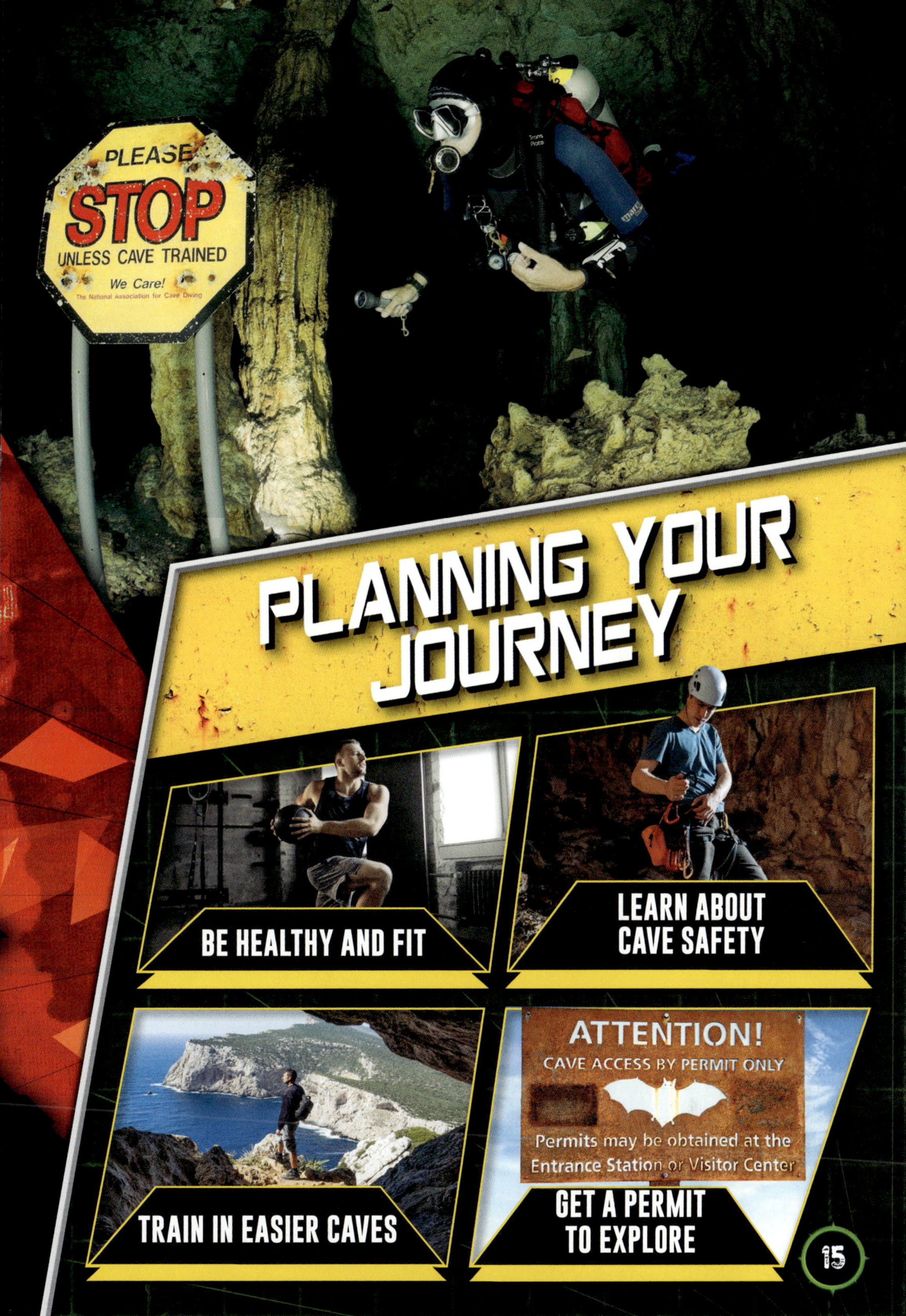

GOING DEEP

Cave explorers need many tools. Most bring two or three lights to see in the dark. Warm, waterproof clothes are also important.

Underwater cave explorers need scuba gear such as air tanks and flippers. They must bring enough air to leave the cave safely. Ice cave explorers wear **crampons** so they do not slip.

Cave explorers face many dangers. Rain aboveground can cause **flash floods** in caves. Explorers must watch the weather.

BEWARE OF ICE

ICE CAVES COLLAPSE EASILY. IF THEY GET TOO WARM, THE ICE FALLS APART!

Caves can also **collapse**. Passages could be blocked. Explorers often wear helmets to avoid falling rocks. They also use tools to find harmful gases such as **carbon monoxide**.

WHAT HAPPENS WHEN YOU BREATHE CARBON MONOXIDE?

DIZZINESS AND SHORTNESS OF BREATH

HEADACHE

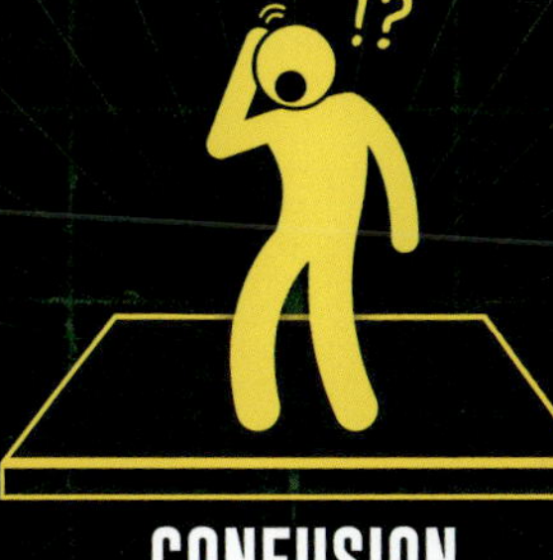

CONFUSION

PASS OUT

Caves can be dangerous to explore. But exploring them can be rewarding. Explorers often make new discoveries. They find things that have never been seen before!

CAVE OF THE CRYSTALS

CAVE OF THE CRYSTALS IS AN UNDERWATER CAVE IN MEXICO. ITS WATER WAS PUMPED OUT FOR MINING. IN 2000, MINERS DISCOVERED HUGE CRYSTALS IN THE CAVE. SOME WERE OVER 30 FEET (9 METERS) LONG!

CAVE OF THE CRYSTALS

They make maps or study how caves formed. They collect **samples** to study cave life. Exploring caves is both fun and good for science!

GLOSSARY

carbon monoxide—a gas with no color or smell that is dangerous to breathe in

caverns—large open spaces within a cave

certified—officially able to do something

collapse—to fall down or cave in

continent—one of Earth's seven large land masses

crampons—pieces of metal with sharp spikes that attach to shoes or boots to help climbers stand on ice

echoes—sounds that repeat over and over after they bounce off a surface

flash floods—sudden floods, usually caused by heavy rain

fragile—easily damaged

insects—small animals with six legs and hard outer bodies; an insect's body is divided into three parts.

lava—hot melted rock that flows out of the earth and has cooled

passages—small areas that people can move through

permits—official papers that show people are allowed to do or have something

samples—small amounts of things that give information about where they were taken from

scuba diving—a type of diving where divers use special breathing equipment that lets them stay underwater for a long time

TO LEARN MORE

AT THE LIBRARY

Castro, Rachel. *Amazing Caves Around the World.* North Mankato, Minn.: Capstone Press, 2019.

Green, Sara. *Caves.* Minneapolis, Minn.: Bellwether Media, 2022.

Huddleston, Emma. *Looking Into Caves.* Mankato, Minn.: The Child's World, 2020.

ON THE WEB

FACTSURFER

Factsurfer.com gives you a safe, fun way to find more information.

1. Go to www.factsurfer.com
2. Enter "exploring caves" into the search box and click 🔍.
3. Select your book cover to see a list of related content.

INDEX

The images in this book are reproduced through the courtesy of: Mark Agnor, front cover (hero); pedrosala, front cover (small explorer); SV Production, front cover (background), pp. 3, 23; wjarek, front cover (stalactites); Gilitukha, p. 4; Dmytro Gilitukha, p. 5; Posnov/ Getty Images, p. 6; Thampitakkull Jakkree, p. 7; Ivan Kurmyshov, p. 8; CHC3537, p. 9; Samantha Nundy/ Alamy, p. 10; Nature Picture Library/ Alamy, p. 11; Melissa Renwick/ Getty Images, p. 11 (Jill Heinerth); VidEst, p. 11 (stalactites); Pedro Antonio Salaverría Calahorra/ Alamy, p. 12; Buzz Pictures/ Alamy, p. 13; Jellyman Photography, p. 14; Reinhard Dirscherl/ Alamy, p. 15; Syda Productions, p. 15 (healthy and fit); africa_pink, p. 15 (cave safety); Gabriele Maltinti, p. 15 (training); JortPics, p. 15 (permit); Karkhut, p. 16; Roman Mikhailiuk, p. 17; Rowan Romeyn/ Alamy, p. 18; Leremy, p. 19 (icons); Alexander Van Driessche/ Wikipedia, p. 20; Robbie Shone/ Alamy, p. 21.